IOANNIS VASILEIOU

THE EUROPEAN
UNION EXPANSION
INTO SPACE

ATHENS 2018

PUBLISHED IN GREEK BY HISTORICAL QUEST IN 2017

TRANSLATED INTO ENGLISH IN 2018

TRANSLATED BY THE AUTHOR HIMSELF

ART DIRECTOR: SOFIA LIVIERATOU

IOANNIS VASILEIOU

ACKNOWLEDGEMENTS

To my beloved Maria Nikou for her encouragement and moral support

TABLE OF CONTENTS

INTRODUCTION .... page 05

CHAPTER 1: THE FOUR MAIN SECTORS OF THE EUROPEAN UNION SPACE POLICY .... page 13

CHAPTER 2: THE PRACTICE OF THE EUROPEAN UNION SPACE POLICY: PRINCIPAL INSTITUTIONS AND AGENCIES .... page 38

CHAPTER 3: CONCLUDING REMARKS .... page 62

BIBLIOGRAPHY .... page 68

INTRODUCTION

The European Union (EU) expansion into space, combined with the promotion of scientific space research in general, have been not only desirable, but also indispensable objectives for several decades, perhaps even from the very first day of the Union's establishment.

In the past, numerous difficulties and risks, coupled with the lack of sufficient knowledge and infrastructure, made this tremendously ambitious target outstandingly difficult. Even so, more than a few salient issues that have arisen over the years, namely the necessity for more accurate weather forecasts, as well as new inventions such as television have made further space exploration mandatory.

Since the 1950s and especially during the second half of the decade, when more systematic efforts for space exploration began to take place, both the scientific interest and the curiosity of ordinary people regarding what exists outside planet Earth peaked. It was perfectly normal that at that time no one could realize the usefulness of space in the context of everyday life and everyday problems.

And of course, nobody could have imagined that, at some point, space would be at the service of human society in order for a vast array of leading global challenges, namely climate change, environmental protection, energy management, health problems, "burning" issues in the context of defense and security, industrial competitiveness, unemployment, disaster relief, crises and emergencies to be adequately addressed.

Over time, space has ceased to appear as a dangerous abyss and has actually turned into a dimension, the rational exploitation of which could serve plentiful needs and actively contribute to a notable improvement regarding quality of life.

We avow that space exploitation has already had an exceptionally positive worldwide impact, since it has managed to successfully provide solutions to problems that in other cases either could not be solved, or their treatment would be exceedingly complicated and protracted.

Nowadays, the concept of space has practically become an integral part of day-to-day discussions as abundant foremost sectors, namely telecommunications and international financial systems, are indeed based on space technologies (ASD-EUROSPACE, 2016; Europa/Space, 2017; EU4Journalists, 2011; Procaccia and Sido, 2012; Vereshchetin, 2010).

In other words, the entire globe currently depends heavily on space, and, hence, scientists worldwide have been incessantly working in order to acquire further knowledge that is highly likely to lead to new discoveries and inventions.

It is considered utopian and perhaps a bit arrogant for someone to state that he possesses deep knowledge of space, since its vastness only provokes awe, but the essence is that, thus far, research has performed gigantic steps towards a satisfactory awareness of dozens of fundamental aspects and components (Europa/Space, 2017; EU4Journalists, 2011; Procaccia and Sido, 2012).

It goes without saying that space research has indubitably been evolving at the speed of light and the Union has already invested heavily in it, always taking the overriding interest of its citizens into careful consideration.

Following a meticulous scrutiny of numerous EU archives, we have drawn the conclusion that, in general, the total cost of space programs is far from insignificant and, therefore, virtually impossible to be covered by individual European countries (Europa/Space, 2017; EU4Journalists, 2011; Procaccia and Sido, 2012).

While referring to current or past "superpowers", namely the United States, or the former Soviet Union, things are in some cases changing due to the fact that financial figures differ appreciably. Even in such cases, though, a number of space operations prove to be economically unprofitable.

Within EU member states, the cost of implementing space programs, or at least the majority of them, is considered to be excessive. Hence, according to several EU experts, numerous European countries have thus far succeeded in effectively pooling both their technological and financial resources. Indisputably, such an achievement is of vital importance. Their precise target is space policy management in the most resourceful manner via the European Commission and in close cooperation with the European Space Agency (ESA) (Belgian Science Policy, 2017; Blau, 2011; Dempsey, 2003; Europa/Space, 2017; EU4Journalists, 2011; Krige and Russo, 2000; Marta, 2013; Masson-Zwaan, 2010; Procaccia and Sido, 2012; Robinson and Romancov, 2014).

The Union's space policy can incontrovertibly be characterized as a titanic project and, concurrently, a predominantly fascinating challenge towards discovering new paths of progress and growth (Commission of the European Communities, 2003; Europa/Space, 2017; European Commission, 2014; EU4Journalists, 2011; Procaccia and Sido, 2012; Vasileiou 2013a, 2013b, 2014a, 2014b, 2015, 2017a, 2017b, 2017c, 2017d and 2018; Vereshchetin, 2010).

Legislation is laid down in Article 189 (Space Policy) of the Treaty on the Functioning of the EU. Emphasis is placed on the fact that Europe is exporting state-of-the-art space systems not only for scientific, but also for commercial purposes. This combination offers the Union a brilliant opportunity to consolidate its global leadership in the context of the always competitive sector of space research and policy (Commission of the European Communities, 2003; Europa/Space, 2017; European Commission, 2014; EU4Journalists, 2011; Procaccia and Sido, 2012; Vasileiou 2013a, 2013b, 2014a, 2014b, 2015, 2017a, 2017b, 2017c, 2017d and 2018; Vereshchetin, 2010).

The current book seeks to painstakingly examine the entire spectrum of factors, elements and parameters that link the Union to space. Programs, actions, processes, policies and strategies have been scrupulously analyzed, while the ultimate goals are a) to draw a number of fruitful conclusions regarding whether space policy has indeed benefited EU citizens and b) to clearly identify future prospects and targets.

Any accurate prediction regarding future developments is absolutely impossible, but the Union on the basis of current data is unremittingly operating (through intensified scientific studies) towards a particularly constructive designing regarding future policies and strategies.

We irrefutably hope for positive future prospects, but the latter will solely be observed over time. Even so, we indisputably acknowledge that the EU has already performed monumental strides in the truly thorny path of space policy and exploration.

Under no circumstances, though, should the still existing economic and financial crisis, as well as the severe (in a number of EU member states) sociopolitical instability be forgotten or ignored, due to the fact that they both inevitably impede initiatives in practically every field of politics.

Examining the subject in general, and not solely from the space policy perspective, the EU is currently being shocked by considerably harsh political and economic tribulations, namely Brexit and the nightmarish Greek economic and financial crisis. We have to admit, though, that despite such anguish problems, the Union is indisputably making unremitting efforts towards a more optimistic future.

This book comprises three chapters. In Chapter 1, a critical analysis of *"The Four Main Sectors of The European Union Space Policy"*, namely Copernicus Earth observation system, Galileo and EGNOS[1] satellite programs, space exploration and space research takes place (Blau, 2011; Dempsey, 2003; Europa/Space, 2017; European Commission, 2014; EU4Journalists, 2011; Marta, 2013; Robinson and Romancov, 2014; Vasileiou 2013a, 2013b, 2014a, 2014b, 2015, 2017a, 2017b, 2017c, 2017d and 2018).

The entire spectrum of the abovementioned sectors has been conscientiously examined in order for the reader to be able to familiarize himself with the entire range of their functions, peculiarities, specificities and future prospects, which are of vital magnitude both for the Union and the rest of the globe.

Chapter 2, entitled *"The Practice of The European Union Space Policy: Principal Institutions and Agencies"*, seeks to investigate the multifarious roles of the European Parliament, the Council of the EU, the European Commission, the European Economic and Social Committee, the Committee of the Regions, the ESA, the GSA[2] [which is the Agency of the European Global Navigation Satellite System (GNSS)], the European Environment Agency (EEA) and the "EU Satellite Centre" (Blau, 2011; Dempsey, 2003; Europa/Space, 2017;

---
1. European Geostationary Navigation Overlay Service.
2. European Global Navigation Satellite Systems Agency.

European Commission, 2017; Krige and Russo, 2000; Marta, 2013; Masson-Zwaan, 2010; Robinson and Romancov, 2014; SatCen, 2017; Vasileiou 2013a, 2013b, 2014a, 2014b, 2015, 2017a, 2017b, 2017c, 2017d and 2018; Vereshchetin, 2010).

It is not difficult to conclude that a detailed examination of the responsibilities and the interdependence of all aforementioned bodies is particularly complex. Consequently, we have been striving to provide a lucid picture in an explicit way, so as to offer the reader adequate knowledge without, nonetheless, tiring or complicating him.

We remain optimistic that the bibliography is sufficient enough for deeper understanding and in order to set light to any dark points that may arise. The EU space policy issue may well be regarded as inexhaustible, hence proper and valid information indubitably requires a systematic monitoring of developments.

At this point, it is essential to emphasize a key parameter, concerning the book's structure. Both in introduction and Chapter 1, a number of functions of key institutions and agencies, systematically analyzed in Chapter 2, have been summarized.

In Chapter 1, we opted for examining the foremost areas of EU space policy and in the following one for proceeding to a more rigorous assessment of the exact role of the Union's key bodies.

Such a decision stems from the fact that we preferred to firstly refer to the four core areas of space policy (so as to provide the reader with a comprehensible general idea regarding the space policy concept) and afterwards to further delve into the (considerably more complicated) tasks of EU institutes.

Finally, in Chapter 3, the main *"Concluding Remarks"* have been drawn, in a systematic attempt to ascertain whether EU space policy can be hitherto characterized as successful.

Our target is to estimate the exact degree of success of strategies and actions to date, while simultaneously referring to the most pivotal future prospects. It is unambiguous though that accurate forecasting regarding the precise future impact is utterly impossible, given the rapid alterations in developments that ceaselessly take place.

We strongly believe that this book will manage to efficaciously contribute to the unremitting scientific research related to the space sector exploration and the systematic promotion of knowledge in terms of the entire array of relevant strategies.

Space policy is widely regarded as the greatest hope for our planet's future. On this logic, in the context of the following chapters, we methodically attempt to justify such a claim via the rational use of concrete arguments.

CHAPTER 1

THE FOUR MAIN SECTORS OF THE EUROPEAN UNION SPACE POLICY

The Union's space policy has always been characterized by an enormously wide range of actions. Complicatedness is more than obvious, but should we wish to summarize, its foremost sectors are the following four: a) Copernicus system, b) Galileo and EGNOS satellite programs, c) space exploration, and d) space research (Blau, 2011; Dempsey, 2003; Europa/Space, 2017; European Commission, 2014; EU4Journalists, 2011; Marta, 2013; Procaccia and Sido, 2012; Robinson and Romancov, 2014; Vasileiou 2013a, 2013b, 2014a, 2014b, 2015, 2017a, 2017b, 2017c, 2017d and 2018).

Perhaps, at first glance, the latter two seem to be somewhat indistinct, but our hitherto conducted research leads us to the conclusion that the EU has indeed taken care of highly specialized policies with a view to a speedy accomplishment of various tangible objectives.

Technological progress, accompanied by the rational use of space data, can for all intents and purposes serve as a fundamental step for the successful achievement of the coveted sustainable development.

Quality research and development, combined with an apposite management of space data, signals and applications, can irrefutably serve as priceless tools for rapidly addressing several "burning" social challenges, namely climate change, health, unemployment, defense, security, energy management, meteorology, monitoring of the atmosphere and the marine environment, environmental protection, management of natural resources, industrial competition, aviation, maritime navigation, railways, agriculture, biotechnology, recycling, forestry, disaster relief, risk signals, rescue operations, civil protection, crises or emergencies, traffic transport, urban planning and tourism.

Our analysis begins with Copernicus system, which focuses on the particularly neuralgic field of Earth observation. The latter can indubitably be characterized as a tremendously complicated process, requiring "surgical precision" actions in terms of both manipulation and coordination.

Copernicus is considered to be the most ambitious (and probably the most systematic) civil Earth observation program. Our research leads us to the assumption that it indubitably has the potential to provide abundant noteworthy benefits (Copernicus, 2017; Europa/Space, 2017; EU4Journalists, 2011; Marta, 2013; Procaccia and Sido, 2012; Robinson and Romancov, 2014; The Hague Manifesto, 2016).

Therefore, it would be utterly wrong to be characterized as "just another simple (and perhaps insignificant) system". On the contrary, it is a set of numerous multifaceted formats designed to resourcefully gather Earth data through satellites and special sensors on the ground, at sea and in the sky (Copernicus, 2017; Europa/Space, 2017; EU4Journalists, 2011; Marta, 2013; Procaccia and Sido, 2012; Robinson and Romancov, 2014; The Hague Manifesto, 2016).

Copernicus provides businesses, policy-makers and the public with a wealth of vital information regarding alterations that take place in the planet and its climate. We acknowledge that, especially nowadays where climate change is a real scourge, any information of this kind literally deserves gold (ASD-EUROSPACE, 2016; Copernicus, 2017; Europa/Space, 2017; EU4Journalists, 2011; Marta, 2013; Masson-Zwaan, 2010; Robinson and Romancov, 2014; The Hague Manifesto, 2016).

We have thus far realized that a systematic collection of such data will indisputably give a massive boost towards the accurate forecasting of future climate trends. Moreover, Copernicus data can be efficaciously applied in the following areas: a) health, b) environmental protection, c) agriculture and forestry, d) disaster response, e) transport, f) urban planning, and g) tourism (ASD-EUROSPACE, 2016; Copernicus, 2017; Europa/Space, 2017; EU4Journalists, 2011; Marta, 2013; Masson-Zwaan, 2010; Procaccia and Sido, 2012; Robinson and Romancov, 2014; The Hague Manifesto, 2016).

It is worth mentioning that the European Commission has been entrusted with both the management and coordination of

Copernicus. The ESA is responsible for satellite infrastructures, while the EEA and individual EU member states are in charge of the sensors (Belgian Science Policy, 2017; Blau, 2011; Commission of the European Communities, 2003; Copernicus, 2017; Europa/Space, 2017; EU4Journalists, 2011; Krige and Russo, 2000; Marta, 2013; Masson-Zwaan, 2010; Robinson and Romancov, 2014).

We remain perfectly sure that, following loads of hard struggles, the Union has effectively been positioned at the forefront in terms of Earth observation and is literally obliged to remain there by incessantly improving both infrastructure and services.

In order to critically delve into the context of Copernicus' analysis, it is essential to highlight the pivotal significance of EUMETSAT, which is in charge of the (irrefutably crucial) meteorological data collected by satellite.

More specifically, EUMETSAT will actively support the Copernicus services responsible for the systematic monitoring of the atmosphere, climate change and the marine environment. The 30 EUMETSAT member countries are a) all EU member states[3] with the exception of Cyprus and Malta, b) Iceland, c) Norway, d) Switzerland and e) Turkey. Additionally, Serbia has been a Cooperating State since 2009 (Belgian Science Policy, 2017; Copernicus, 2017; Europa/Space, 2017; Vasileiou, 2014b and 2017c).

We strongly believe that the aforementioned components

---

3. *We still regard the United Kingdom as an EU member state.*

(meteorological data and monitoring of the atmosphere, climate change and the marine environment) are outstandingly critical for both the present and the future of our heavily damaged planet. Consequently, any actual progress leading to their rational observation results in an immediate positive worldwide impact.

Meteorological data may well serve as a first-class measure for the satisfactory prevention of major disasters. Regrettably, the devastating problem of climate change, accompanied by its plentiful abnormal (and often dangerous) alterations in terms of climatic and meteorological conditions, results in extreme weather phenomena, namely storms, droughts, heat waves and floods.

The safest possible prediction for the longest period of time is the principal goal of meteorology. Consequently, any innovative scientific methods with the capacity of lengthening the period of safe prediction will irrefutably perform colossal work.

A methodical observation of atmospheric pressure and barometric stress can also be regarded as a dominant element concerning the impact of weather phenomena, due to the fact that they are both directly linked to winds and cyclones.

Furthermore, in quite a few cases, increasing air pollution causes not only incurable diseases, but also global warming, which, alas, is currently considered by numerous scientists to be irreversible. Therefore, an incessant monitoring of the atmosphere's composition, particulate matter, dust and greenhouse gas emissions is compulsory.

Likewise, marine environment pollution leads with mathematical precision to the overall destruction of ecosystems and biological resources. Consequently, we have thus far acknowledged that an uninterrupted monitoring of the atmosphere, climate change and the marine environment is absolutely vital for any form of life on Earth and must be meticulously supported by state-of-the-art systems.

Incontrovertibly, the Union has always attached tremendous importance to the encouragement of research and innovation. Such a component indubitably contributes to the rapid development of new scientific and technological methods for more optimistic results and visible progress.

The second space policy sector is Galileo and EGNOS satellite programs. Galileo is the Union's GNSS and we argue that it is the European version of the Russian GLONASS and the US Global Positioning System (GPS) (Blau, 2011; Dempsey, 2003; Europa/Space, 2017; European Commission, 2014; EU4Journalists, 2011; Galileo, 2017; GNSS, 2016; GSA/About GSA, 2015; GSA/What we do, 2016; Marta, 2013; Robinson and Romancov, 2014; The Hague Manifesto, 2016).

We are all familiar with the massive noteworthiness, as well as the broad scope of GPS, so the mere thought of a European version of it, which is likely to be technologically superior, gives the EU unparalleled authority and enhances its status within the international scientific field.

Galileo is considered to be the first civilian-run satellite navigation system and special emphasis must be placed on the fact that it will be indeed compatible with Russian and American systems, but will remain independent of them. This detail should not under any circumstances go unnoticed. A further attention-grabbing finding is the fact that Galileo will be far more accurate and reliable compared to GPS, thanks to its real-time positioning of one meter or less (Blau, 2011; Dempsey, 2003; Europa/Space, 2017; European Commission, 2014; EU4Journalists, 2011; Galileo, 2017; GNSS, 2016; GSA/About GSA, 2015; GSA/What we do, 2016; Marta, 2013; Procaccia and Sido, 2012; Robinson and Romancov, 2014; The Hague Manifesto, 2016).

This brilliant feature truly illustrates the mammoth EU progress with regard to the aforesaid technologies. Perhaps in earlier times it sounded inconceivable that the EU could create satellite systems technologically comparable, or even superior to the two traditional "superpowers", but as we can witness today, such a masterpiece is indeed feasible.

Without exaggeration, the 15th of December 2016 can be regarded as a landmark date for EU satellite technologies. That particular day the Declaration of Initial Services practically gave the green light for the long-awaited start of the Galileo operational phase (Blau, 2011; Dempsey, 2003; Europa/Space, 2017; European Commission, 2014; Galileo, 2017; Marta, 2013; Robinson and Romancov, 2014; The Hague Manifesto, 2016).

A little earlier, on November 17, 2016, four Galileo satellites had been successfully launched by the European launcher Ariane-5, which was modified by specific methods and will solely be used for Galileo. This very launch increased Galileo's formation in 18 satellites (Blau, 2011; Dempsey, 2003; Europa/Space, 2017; European Commission, 2014; Galileo, 2017; Marta, 2013; Robinson and Romancov, 2014; The Hague Manifesto, 2016).

As a matter of fact, this launch has been a milestone in terms of the Union's space history, due to the fact that for the first time four European satellites were lofted into orbit by a single European launcher. Following an exhaustive research in the context of the Union's space policy, we understand that autonomous access to space is a cornerstone in the context of EU Space Strategy (Blau, 2011; Dempsey, 2003; Europa/Space, 2017; European Commission, 2014; EU4Journalists, 2011; Galileo, 2017; Marta, 2013; Robinson and Romancov, 2014; The Hague Manifesto, 2016).

An additional interesting feature is that Galileo satellites were named after children who had won an EU-wide drawing competition. The four satellites are called Lisa (Hungary), Tijmen (Netherlands), Kimberley (Malta) and Antonianna (Italy) (Europa/Space, 2017; European Commission, 2014; Galileo, 2017; Marta, 2013).

The issue calling for further analysis, though, is why the Union stands in need for Galileo, although we assume in advance that the reader has already guessed the answer.

Nowadays, it is crystal clear that merciless competition takes place in terms of space policies and satellite systems. Space has eventually turned into an immense arena where interests at stake are enormous.

The EU must be capable of successfully coping with other major powers (and if possible, overcome them) in the context of technological and scientific achievements, both in terms of strengthening its prestige on a global scale and for the best possible service of its citizens.

Therefore, it is imperative for Galileo to be in position to satisfactorily compete with miscellaneous equally noteworthy systems[4]. Moreover, a sagacious development of space-related technologies and satellites is a stepping stone for industrial competitiveness in general (Blau, 2011; Dempsey, 2003; Europa/Space, 2017; European Commission, 2014; Galileo, 2017; GNSS, 2016; GSA/About GSA, 2015; GSA/What we do, 2016; Marta, 2013; Procaccia and Sido, 2012; Vasileiou 2013a, 2013b, 2014a, 2014b, 2015, 2017a, 2017b, 2017c, 2017d and 2018).

According to the Union's archives, as well as additional relevant independent studies, it is expected that over the first 20 years of operations, Galileo will deliver approximately €90 bn to the EU economy. Such a feature must not be ignored, since it can lead to various interesting conclusions regarding the future economic situation of the Union, which everybody hopes to be less problematic (Blau, 2011; Dempsey, 2003; Europa/Space, 2017; European Commission, 2014; Galileo, 2017; GNSS, 2016; GSA/About GSA, 2015; GSA/What we do, 2016; Marta, 2013;

---

4. Such as the aforementioned Russian GLONASS, the Chinese BeiDou and the new generation of an upgraded, technologically improved US GPS.

Procaccia and Sido, 2012; Vasileiou 2013a, 2013b, 2014a, 2014b, 2015, 2017a, 2017b, 2017c, 2017d and 2018).

Galileo is owned and funded by the Union. The Commission is in charge of the overall responsibility for the program, the management and the supervision of implementation regarding the entire range of activities on behalf of the Union. Concomitantly, the design of Galileo, its deployment, the development of the new generation of systems and the technical development of infrastructure have been assigned to the ESA (Blau, 2011; Commission of the European Communities, 2003; Dempsey, 2003; Europa/Space, 2017; European Commission, 2014; EU4Journalists, 2011; Galileo, 2017; GNSS, 2016; GSA/About GSA, 2015; GSA/What we do, 2016; Krige and Russo, 2000; Marta, 2013; Masson-Zwaan, 2010).

The Commission has entrusted the GSA with the program's operational management. This actually means that the latter is required to ensure that Galileo services are ceaselessly provided, without tribulations and based on accurate planning (Blau, 2011; Commission of the European Communities, 2003; Dempsey, 2003; Europa/Space, 2017; European Commission, 2014; EU4Journalists, 2011; Galileo, 2017; GNSS, 2016; GSA/About GSA, 2015; GSA/What we do, 2016; Krige and Russo, 2000; Marta, 2013; Masson-Zwaan, 2010).

Galileo is expected to be completed by 2020 and will include 24 satellites and six active back-ups. At present, as it has already been noted, it consists of a formation of 18 satellites and we highlight the fact that with these orbiting satellites and their supporting ground infrastructure, Galileo, following an extensive period of scrutiny and testing, currently offers three pivotal "Initial Services" (ASD-EUROSPACE, 2016; Belgian Science Policy, 2017; Blau, 2011; Commission of the European Communities, 2003; Dempsey, 2003; Europa/Space, 2017; European Commission, 2014; EU4Journalists, 2011; Galileo, 2017; GNSS, 2016; GSA/About GSA, 2015; GSA/What we do, 2016; Krige and Russo, 2000; Marta, 2013; Masson-Zwaan, 2010; Robinson and Romancov, 2014).

The first is the "Open Service", which can effectively be characterized as a free "mass-market" service for positioning, timing and navigation. The second is the "Public Regulated Service" for government-authorized users, namely the police, customs officers and civil protection services. Attention has to be paid to the fact that this system is sufficiently robust and fully encrypted, with the aim of providing a sufficient service continuity to government users in terms of crisis situations, or emergencies (Blau, 2011; Dempsey, 2003; Europa/Space, 2017; European Commission, 2014; Galileo, 2017; GNSS, 2016; GSA/About GSA, 2015; GSA/What we do, 2016; Marta, 2013).

The third and final one is the "Search and Rescue Service", which is considered to be the European contribution to COSPAS-SARSAT, an international organization which detects and accurately locates radio beacons activated by individuals, planes or ships in danger. Galileo data successfully provide significant assistance in locating beacons and rescuing people in distress in practically every kind of environment. Such a capacity is irrefutably crucial for numerous important reasons (Blau, 2011; Dempsey, 2003; Europa/Space, 2017; European Commission, 2014; Galileo, 2017; GNSS, 2016; GSA/About GSA, 2015; GSA/What we do, 2016; Marta, 2013; Robinson and Romancov, 2014; Vasileiou 2013a, 2013b, 2014a, 2014b, 2015, 2017a, 2017b, 2017c, 2017d and 2018).

The abovementioned services are free of charge and available to businesses, authorities, and citizens alike. Galileo's "Initial Services" are being managed by the GSA, which is based in Prague (Blau, 2011; Dempsey, 2003; Europa/Space, 2017; European Commission, 2014; Galileo, 2017; GNSS, 2016; GSA/About GSA, 2015; GSA/What we do, 2016; Marta, 2013; Robinson and Romancov, 2014).

Galileo started providing services in 2016, and its potential applications include a) farming, b) rescue operations, c) civil protection, d) traffic and transport management and e) time stamping and time synchronization (ASD-EUROSPACE, 2016; Blau, 2011; Dempsey, 2003; Europa/Space, 2017; European Commission, 2014; Galileo, 2017; GNSS, 2016; GSA/About GSA, 2015; GSA/What we do, 2016; Marta, 2013; Robinson and Romancov, 2014).

Emphasis must also be placed on Galileo partners, which are three. The first is the European Commission, in charge of both the management and the full funding of the program. The second partner is the ESA, responsible for the development, the design, the procurement and the validation, tasks of utmost importance (Belgian Science Policy, 2017; Blau, 2011; Commission of the European Communities, 2003; Dempsey, 2003; Europa/Space, 2017; European Commission, 2014; EU4Journalists, 2011; Galileo, 2017; GNSS, 2016; GSA/About GSA, 2015; GSA/What we do, 2016; Krige and Russo, 2000; Marta, 2013; Masson-Zwaan, 2010; Procaccia and Sido, 2012).

It is worth noting that the ESA has already partly funded the definition, development and in-orbit validation phases of Galileo, which is a apparent indication of its caliber. The third partner is the European GNSS Agency, entrusted with the responsibility of operations as soon as the system is completed (Belgian Science Policy, 2017; Blau, 2011; Commission of the European Communities, 2003; Dempsey, 2003; Europa/Space, 2017; European Commission, 2014; EU4Journalists, 2011; Galileo, 2017; GNSS, 2016; GSA/About GSA, 2015; GSA/What we do, 2016; Krige and Russo, 2000; Marta, 2013; Masson-Zwaan, 2010; Procaccia and Sido, 2012).

A further principal target in the context of EU investments in Galileo is the creation of new jobs and market opportunities in terms of the particularly interesting as well as highly specialized sectors of satellite signal receivers and satellite-based applications (Blau, 2011; Dempsey, 2003; Europa/Space, 2017; European Commission, 2014; EU4Journalists, 2011; Galileo, 2017; GNSS, 2016; GSA/About GSA, 2015; GSA/What we do,

2016; Marta, 2013; Procaccia and Sido, 2012; Robinson and Romancov, 2014; Vasileiou 2013a, 2013b, 2014a, 2014b, 2015, 2017a, 2017b, 2017c, 2017d and 2018).

We are actually pleased to witness that within the context of space policy, combating unemployment, particularly among young people, remains a paramount concern of the Union.

Such an assumption provides us with concrete evidence of the fact that a strong and united Europe unquestionably has the potential to generously offer significant opportunities to the community for the gradual improvement of the quality of life, even in periods of intense crises and generalized instability.

EGNOS is a satellite system that improves GPS accuracy to within two meters (95%) instead of 10 that the latter normally provides. Such a detail is tremendously crucial, due to the fact that this way GPS becomes infinitely safer for both aviation and maritime navigation (EGNOS, 2017; Europa/Space, 2017; Galileo, 2017; GNSS, 2016; GSA/About GSA, 2015; GSA/What we do, 2016; Procaccia and Sido, 2012; Robinson and Romancov, 2014).

Over and above that, it possesses the ability to alert users in time for possible problems concerning GPS signals and has been widely regarded as the precursor to Galileo. On top of that, we underline the fact that EGNOS is Europe's first systematic attempt in terms of satellite navigation and is considered to be a "Satellite-Based Augmentation System (SBAS)" with the aim of substantially improving the accuracy of basic satellite navigation signals in Europe (EGNOS, 2017; Europa/Space, 2017; Galileo,

2017; GNSS, 2016; GSA/About GSA, 2015; GSA/What we do, 2016; Procaccia and Sido, 2012; Robinson and Romancov, 2014).

EGNOS uses three satellites so as to satisfactorily correct GPS errors, while it is a common assumption that it provides far more accurate positioning data. EGNOS practically consists of transponders installed on three geostationary satellites and a multifunctional interconnected ground network of approximately 40 positioning stations and four mission control centers. Special attention must be paid to the fact that, according always to the Union, EGNOS possesses the appropriate technical capacity in order to extend to both Middle East and North Africa (EGNOS, 2017; Europa/Space, 2017; Galileo, 2017; GNSS, 2016; GSA/About GSA, 2015; GSA/What we do, 2016; Robinson and Romancov, 2014).

Contrary to Galileo, EGNOS is pan-European rather than global and depends on GPS. Furthermore, it is essential to mention that it is a joint project of the European Commission, the ESA and Eurocontrol, which is the "European Organisation for the Safety of Air Navigation". The latter comprises 41 member countries (all EU member states, Switzerland, Turkey, Norway, FYROM, Monaco, the Republic of Moldova, Albania, Bosnia and Herzegovina, Ukraine, Serbia, Montenegro, Armenia and Georgia) (Belgian Science Policy, 2017; Commission of the European Communities, 2003; EGNOS, 2017; Europa/Space, 2017; Galileo, 2017; Krige and Russo, 2000; Marta, 2013; Masson-Zwaan, 2010; Robinson and Romancov, 2014; The Hague Manifesto, 2016).

The ESA used to be in charge of managing EGNOS' development, under a tripartite agreement with the Commission and Eurocontrol. The ownership of EGNOS assets was successfully transferred from the ESA to the Commission in April 2009 and EGNOS officially entered service on October 1, 2009. Through a contract with the GSA, the service is delivered by ESSP SaS, which is the European Satellite Services Provider (Commission of the European Communities, 2003; Dempsey, 2003; EGNOS, 2017; Europa/Space, 2017; GNSS, 2016; GSA/About GSA, 2015; GSA/What we do, 2016; Krige and Russo, 2000; Marta, 2013; Masson-Zwaan, 2010; Procaccia and Sido, 2012; Robinson and Romancov, 2014; The Hague Manifesto, 2016).

Since 2014, the GSA has successfully been the EGNOS program manager under delegation from the Commission and it has to be noted that the ESA is the design and procurement agent working on behalf of the Commission (Commission of the European Communities, 2003; Dempsey, 2003; EGNOS, 2017; Europa/Space, 2017; GNSS, 2016; GSA/About GSA, 2015; GSA/What we do, 2016; Krige and Russo, 2000; Marta, 2013; Masson-Zwaan, 2010; Procaccia and Sido, 2012; Robinson and Romancov, 2014; The Hague Manifesto, 2016). According to numerous EU experts, EGNOS has thus far been characterized by an increasingly serviceable functionality. Such a remarkable feature literally obliges us to comment on a number of its foremost application examples.

First and foremost, it applies to the aviation sector, due to the fact that it can actively contribute to a far more satisfactory navigation and, concomitantly, facilitate the successful design of more efficacious routes (EGNOS, 2017; Europa/Space, 2017; GNSS, 2016; GSA/About GSA, 2015; GSA/What we do, 2016; Robinson and Romancov, 2014).

A further noteworthy application is in the field of traffic control, since it is in position to offer top-class assistance towards the rapid improvement of the enormously important emergency response times. Moreover, its role is absolutely pivotal in the context of precision agriculture, in particular as regards both virtual fencing and the effective detection of cow fertility (EGNOS, 2017; Europa/Space, 2017; GNSS, 2016; GSA/About GSA, 2015; GSA/What we do, 2016; Robinson and Romancov, 2014).

What is more, EGNOS is implemented in the field of search and rescue, due to the fact that it indubitably has the potential to significantly facilitate the landing of helicopters under thorny conditions. Additionally, it is practically indispensable in the context of rail transport, since it provides priceless assistance in order for the exact location of trains to be tracked (EGNOS, 2017; Europa/Space, 2017; GNSS, 2016; GSA/About GSA, 2015; GSA/What we do, 2016; Robinson and Romancov, 2014).

At this instant, a methodical clarification regarding the role of the European Commission in terms of EGNOS seems compulsory for a more coherent understanding. In summary, we assert that the Commission is practically analyzing the impact of EGNOS on competitiveness in the context of four particularly neuralgic parts of the Union's economy (EGNOS, 2017; Europa/Space,

2017; Robinson and Romancov, 2014; Vasileiou 2013a, 2013b, 2014a, 2014b, 2015, 2017a, 2017b, 2017c, 2017d and 2018).

The first part is the precise contribution of the European space industry to the effective construction of global satellite navigation systems. The second relates to the functional provision of services[5], whereas the third part is the European applications industry, which depends a great deal on the provision of services to supply the hardware and software necessary for the astute exploitation of satellite signals. Finally, the fourth part concerns the end users[6] (EGNOS, 2017; Europa/Space, 2017; Robinson and Romancov, 2014).

EGNOS offers three multifunctional high-performance navigation and positioning services, which can be competently used in the context of agriculture, aviation and several other crucial applications. These are the "Open Service", the "Safety-of-Life Service" and the «Commercial Service or 'EGNOS Data Access Service (EDAS)'" (EGNOS, 2017; Europa/Space, 2017; GNSS, 2016; GSA/About GSA, 2015; GSA/What we do, 2016; Robinson and Romancov, 2014).

In general, we presuppose that the EU economy is highly dependent on satellite navigation services in a wide range of applications including transport, telecommunications, logistics

---

5. *More specifically to European businesses supplying commercial or public navigation, positioning or timing services.*
6. *Such as businesses that efficaciously use applications and services provided by satellite signals.*

and energy. A steady growth in the market for satellite navigation services can actually be observed, and it is estimated that by 2022, its value will reach €250 bn a year (EGNOS, 2017; Europa/Space, 2017; Galileo, 2017; Procaccia and Sido, 2012; Robinson and Romancov, 2014; Vasileiou 2013a, 2013b, 2014a, 2014b, 2015, 2017a, 2017b, 2017c, 2017d and 2018).

On top of that, it is worth noting that approximately 6% to 7% of the Union economy currently depends on the availability of global navigation satellite signals. With the valuable aid of precise calculations, such details clearly highlight the tremendous significance of space in purely economic terms (EGNOS, 2017; Europa/Space, 2017; Galileo, 2017; Procaccia and Sido, 2012; Robinson and Romancov, 2014; Vasileiou 2013a, 2013b, 2014a, 2014b, 2015, 2017a, 2017b, 2017c, 2017d and 2018).

Four are the overarching objectives of the Union's satellite navigation programs. The first is the rapid accomplishment of the much-needed technological independence with regard to other related global systems. The second concerns the fastest possible mobilization of both strategic and economic advantages of having European control over the unremitting availability of satellite navigation services (EGNOS, 2017; Europa/Space, 2017; Galileo, 2017; GNSS, 2016; GSA/About GSA, 2015; GSA/What we do, 2016; Marta, 2013; Procaccia and Sido, 2012).

The third one is to sufficiently encourage, support and facilitate the development of new services and products based on satellite signals, while the fourth is to generate related technological benefits for innovation, research and development (EGNOS, 2017; Europa/Space, 2017; Galileo, 2017; GNSS, 2016; GSA/About GSA, 2015; GSA/What we do, 2016; Marta, 2013; Procaccia and Sido, 2012).

In other words, we are indeed witnessing the vast amount of hopes the Union has put on satellite services for a more optimistic present and (hopefully) a more heartening future. It would be utterly imprudent to ignore the fact that any technology based on, or derived from satellites, can resolve more than a few burning problems and give even greater impetus for innovative applications, inventions and discoveries.

It is perfectly reasonable, therefore, for global competition to be relentless within the aforesaid areas. Be that as it may, the Union has so far convinced us that it is unmistakably characterized by the will, as well as the know-how, in order to continue playing a leading role at the international level.

The dynamic combination of Galileo and EGNOS, besides promoting science and knowledge, gives a superior entity to the Union on a global scale. Such an acknowledgement provides us with plentiful aspirations regarding the Union's future progress in general.

The third space policy sector is space exploration, the notability of which can be demonstrated by the fact that it has been widely recognized as a truly decisive element in terms of scientific discovery and technological innovation on issues relating to health, environment, energy management, biotechnology and recycling (ASD-EUROSPACE, 2016; Blau, 2011; Commission of the European Communities, 2003; Europa/Space, 2017; Vasileiou 2013a, 2013b, 2014a, 2014b, 2015, 2017a, 2017b, 2017c, 2017d and 2018).

As it has already been mentioned, the cost of space programs has been reaching skyscraping amounts. Hence, a fruitful international cooperation is indispensable, in order for further tribulations to be avoided.

It is imperative for the EU to remain at the forefront of space exploration and, thus, it is obligatory to regularly update and redesign its planning. The latter must be based not only on international developments, but also on everyday problems within its territory which can decisively affect (mainly economically) its space strategies.

One of the Union's major concerns is the international debate regarding a constructive cooperation in terms of space exploration, mainly with the USA, China and Russia. Such discussions have been considerably critical for several obvious reasons (Blau, 2011; Europa/Space, 2017; EU4Journalists, 2011; Procaccia and Sido, 2012).

The Union actively participates in the International Space Exploration Forum, which is an outstandingly crucial series of meetings at ministerial level aiming at the best possible international coordination in the particularly neuralgic context of space exploration (Blau, 2011; Europa/Space, 2017; EU4Journalists, 2011; Procaccia and Sido, 2012).

Our overall research leads us to the ascertainment that the EU has so far been extremely active in the aforesaid context, and we truly wish to carry on likewise, despite the perpetually growing degree of complicatedness.

The fourth and final sector is space research. It goes without saying that one of the Union's principal goals is the actual development of a competitive independent European space industry. According to the Union, such a pursuit is indeed possible, despite a number of (inevitable) difficulties. If Europe wishes to maintain its successful access to space, focusing on space research and innovation is mandatory (Commission of the European Communities, 2003; Europa/Space, 2017; EU4Journalists, 2011; Procaccia and Sido, 2012).

We argue that whenever it comes to research and innovation within the Union, the first thing that springs to mind is the Horizon 2020 program. It is precisely through the latter that the funding of space research projects under the €13.5 bn Leadership in Enabling and Industrial Technologies (LEIT) work program is available (Dempsey, 2003; Europa/Space, 2017; EU4Journalists, 2011; HORIZON 2020/Leadership in Enabling and Industrial Technologies, 2017; Procaccia and Sido, 2012; Vasileiou 2013a, 2013b, 2014a, 2014b, 2015, 2017a, 2017b, 2017c, 2017d and 2018).

The funding areas are the following five: a) Earth observation, b) applications in Satellite Navigation-Galileo, c) the Small and Medium-sized Enterprises (SMEs) instrument, d) the protection of European assets in space, as well as from space and e) the competitiveness of the European space sector[7] (Dempsey, 2003; Europa/Space, 2017; EU4Journalists, 2011; HORIZON 2020/Leadership in Enabling and Industrial Technologies, 2017; Procaccia and Sido, 2012).

Special attention will be paid to the following areas of LEIT action: A) public-private partnerships, B) research and innovation, with the desirable aim of substantially enhancing European industrial capabilities and business perspectives[8], C) specific contributions towards the rapid resolution of the (often thorny) "Societal Challenges" and the so-called "Focus Areas", D) international cooperation, as well as fruitful research and innovation, E) cross-cutting Key Enabling Technologies and F) the rational seizing of opportunities within the context of the (always important) Information and Communication Technologies (Commission of the European Communities, 2003; HORIZON 2020/Leadership in Enabling and Industrial Technologies, 2017; Procaccia and Sido, 2012; Vasileiou 2013a, 2013b, 2014a, 2014b, 2015, 2017a, 2017b, 2017c, 2017d and 2018).

In the following chapter, the entire spectrum of duties of the European Parliament, the Council of the EU, the European Commission, the European Economic and Social Committee, the Committee of the Regions, the ESA, the GSA, the EEA and

---

7. *Technology and science.*
8. *Including of course SMEs.*

the EU Satellite Center, will be scrupulously analyzed.

Our dominant aim is to conduct as simple as possible a critical scrutiny of this complex system, in order for the reader to be able to effectively delve into issues of both structure and functions in general.

We have thus far realized that space policy is not solely about space itself. On the contrary, it is more about the steadily increasing utility of the latter in successfully addressing plentiful flaming problems that have plagued our planet for many decades.

Therefore, information on space policy is not intended only for astrophysicists, astronomers, aircraft builders, satellite manufacturers and related scientists, but also for every citizen who is confronted with everyday problems and hopes for a better economic situation, a more systematic environmental protection, an effective confrontation of climate change, a more functional transport system, a lower unemployment rate and in general a substantial improvement regarding quality of life.

We take for granted that a non-conflictual Europe is a prerequisite for further progress and is undeniably capable of significantly contributing towards a sufficient raise in terms of the standard of living and education of every citizen, accompanied by the rapid provision of a satisfactory response to dozens of global problems and challenges. We irrefutably support the fact that the EU is one of the most fundamental pillars of growth, progress and stability on an international scale and is relentlessly working towards strengthening its influence in the context of the global economic and sociopolitical sphere.

So far, our assessment regarding the Union's policies (not solely in the context of the space sector) is that the existence, as well as the shrewd operation of a strong and well-organized EU is of vital importance for the absolute benefit of the entire globe.

CHAPTER 2

THE PRACTICE OF THE EUROPEAN UNION SPACE POLICY: PRINCIPAL INSTITUTIONS AND AGENCIES

Speaking about the practice of the Union's space policy, a number of paramount institutions and agencies, namely the European Parliament, the Council of the EU, the European Commission, the European Economic and Social Committee, the Committee of the Regions, the ESA, the GSA, the EEA and the EU Satellite Center, are being actively involved (Belgian Science Policy, 2017; Blau, 2011; Dempsey, 2003; Europa/Space, 2017; European Commission, 2017; Krige and Russo, 2000; Marta, 2013; Masson-Zwaan, 2010; Robinson and Romancov, 2014; SatCen, 2017; Vasileiou 2013a, 2013b, 2014a, 2014b, 2015, 2017a, 2017b, 2017c, 2017d and 2018; Vereshchetin, 2010).

More than a few functions regarding the majority of the abovementioned bodies have already been summarized. The target of this chapter, though, is to discuss their exact role in detail, with a view to a more coherent understanding.

The purpose of this book is not to provide a superficial

reference to the methods of designing and practicing space policy, but to offer the reader the opportunity to judge whether EU strategies are indeed rational both for the promotion of science and the immediate service of the community. The latter can indisputably be characterized as an utterly gigantic mission.

Our analysis begins by examining the Committee on Industry, Research and Energy (ITRE) of the European Parliament. ITRE's overarching aim is the efficacious reindustrialization, which has always been a pivotal EU objective (Commission of the European Communities, 2003; European Commission, 2016; European Parliament, 2017; Procaccia and Sido, 2012; Vasileiou, 2013a, 2014a, 2017a and 2017b).

In order to turn this exceedingly complicated objective into reality, it is indispensable for ITRE to meticulously focus on four critical areas.

The first concerns the urgent stipulation to ensure that both the economic and the legal framework of the Union indeed provide the European industrial sector with the coveted possibility of becoming more innovative. Thus, a rapid and functional adaptation and mobilization in terms of the Union's R&D and education programs is enormously crucial, in order for them to properly respond to precise industrial requirements (Commission of the European Communities, 2003; European Commission, 2016; European Parliament, 2017; Procaccia and Sido, 2012; Vasileiou, 2013a, 2014a, 2017a and 2017b).

We have to admit that such a process is exceedingly complicated. The Union is well aware of this and consequently lots of systematic efforts have effectively begun. According to the view of numerous EU experts, up to now, dozens of positive results have already been observed and chances suggest that the future will be even more heartening.

The second area is SMEs and the sufficient development of their growth and competitiveness. This objective has been widely regarded as one of the most decisive for the entire Union. We stress the fact that the uninterrupted development of the aforementioned enterprises will categorically serve as a key propellant towards the coveted (as well as absolutely necessary) revitalization of the European economy (Commission of the European Communities, 2003; European Commission, 2016; European Parliament, 2017; Vasileiou, 2013a, 2014a, 2017a and 2017b).

The third area concentrates on the fastest possible establishment of a true European digital market, while the fourth and final one concerns the satisfactory exploitation of the full potential of the Union's common energy policy (Commission of the European Communities, 2003; European Parliament, 2017; Vasileiou, 2013a, 2014a, 2017a and 2017b).

The aforesaid fourth area is a cardinal factor regarding EU economy competitiveness in the context of the always prickly international arena. Nowadays, that the global economic and financial crisis has not yet been terminated, such an element proves to be of vital significance in order for further problems to be methodically avoided (Commission of the European Communities, 2003; European Parliament, 2017; Vasileiou, 2013a, 2014a, 2017a and 2017b).

ITRE remains in charge of a wide range of particularly critical sectors. Emphasis on detail, precision and vigilant handling unequivocally illustrate its truly multifarious role (Commission of the European Communities, 2003; European Parliament/ Annex V: Powers and Responsibilities of Standing Committees, 2017; Procaccia and Sido, 2012; Vasileiou, 2013a, 2014a, 2017a and 2017b).

Firstly, ITRE is held responsible for the Union's industrial policy, the entire array of related measures, as well as the rational application of new technologies[9]. Secondly, it is in charge of the European space policy, which is an utterly gargantuan task, characterized by numerous key parameters (Commission of the European Communities, 2003; European Parliament/Annex V: Powers and Responsibilities of Standing Committees, 2017; Procaccia and Sido, 2012; Vasileiou, 2013a, 2014a, 2017a and 2017b).

Thirdly, it has been entrusted with the EU research and innovation policy, which includes science, technology and a well-planned dissemination[10] in terms of research findings. Moreover, ITRE is in charge of a) the Euratom Treaty and the

---

9. *Including specific measures regarding SMEs.*
10. *Accompanied by a serviceable utilisation.*

Euratom Supply Agency and b) nuclear safety, decommissioning and waste disposal in the context of the nuclear sector (Commission of the European Communities, 2003; European Parliament/Annex V: Powers and Responsibilities of Standing Committees, 2017; Procaccia and Sido, 2012; Vasileiou, 2013a, 2014a, 2017a and 2017b).

Over and above that, it is preoccupied with a) the actions of the European Research Council, the "Joint Research Centre", the European Institute of Innovation and Technology and the Institute for Reference Materials and Measurements, b) the International Thermonuclear Experimental Reactor (ITER) and the Joint European Torus (JET) and c) more than a few miscellaneous noteworthy projects within the same area (European Parliament/ Annex V: Powers and Responsibilities of Standing Committees, 2017; Vasileiou, 2013a, 2014a, 2017a and 2017b).

On top of that, it remains responsible for the information society, information technology, as well as communications networks and services. We must not ignore the attention-grabbing fact that a) technologies and security aspects and b) the creation and further development of trans-European networks in the context of the neuralgic sector of telecommunications' infrastructure are actually included there. A further noteworthy ITRE task is to satisfactorily deal with the European Union Agency for Network and Information Security (ENISA) actions (European Parliament/Annex V: Powers and Responsibilities of Standing Committees, 2017; Procaccia and Sido, 2012;

Vasileiou, 2013a, 2014a, 2017a and 2017b).

Finally, ITRE has been entrusted with EU energy policy measures both in general terms and for the establishment and proper operation of the internal energy market, including measures regarding a) the interconnection of energy networks, as well as energy efficiency[11], b) the security of energy supply in the Union and c) energy saving, energy efficiency and the rapid development of new and renewable energy forms (European Parliament/Annex V: Powers and Responsibilities of Standing Committees, 2017; Procaccia and Sido, 2012; Vasileiou, 2013a, 2014a, 2017a and 2017b).

Such a multifaceted grid reveals not only the complexity regarding the entire array of ITRE functions, but also its notability. We cannot disagree with the opinion of numerous experts, arguing that ITRE is one of the Parliament's top committees, with several massively important tasks to successfully carry out.

At this point, the multipart role of the Competitiveness Council (COMPET) will be critically examined. Its principal aim is to directly boost growth and competitiveness within the Union, while its foremost areas are a) space, b) the internal market, c) industry and d) research and innovation (ASD-EUROSPACE, 2016; Commission of the European Communities, 2003; European Commission, 2016; European Council/Council of the European Union, 2017; Procaccia and Sido, 2012; Vasileiou,

---

11. *Including the creation of trans-European networks in terms of the (outstandingly crucial) energy infrastructure sector.*

2013a and 2014a).

Always according to the agenda, the Competitiveness Council brings together Ministers of all member states in charge of space, economy, trade, industry, research and innovation. Additionally, relevant European Commissioners take part in the meetings and it is worth mentioning that meetings take place at least four times during the year (ASD-EUROSPACE, 2016; Commission of the European Communities, 2003; European Commission, 2016; European Council/Council of the European Union, 2017; Procaccia and Sido, 2012; Vasileiou, 2013a and 2014a).

The momentousness of the Competitiveness Council practically obliges us to carry out a more extensive analysis of the four aforesaid policy areas.

As far as a) space and b) research and innovation are concerned, the Council is engaging in a considerably purposeful cooperation with the ESA, while simultaneously ensuring that it interminably promotes the technological and scientific base of European industry in a methodical way[12]. In the context of the internal market, the Council as law-maker aims towards a rapid removal of all barriers that hinder and somehow intercept the smooth cross-border flow of labor, capital, products and services. Finally, speaking about industry, the Council is in favor of a rather satisfactory combination of a horizontal approach[13]

---

12. *More specifically, the Council has been ceaselessly operating with a view to enhancing the international competitiveness of the European industry, combined with the fastest possible achievement of growth and the creation of new jobs.*
13. *Characterized by the overriding consideration of rapidly integrating key industrial policy concerns into all other relevant Union policies.*

with a utilitarian sector-specific one (ASD-EUROSPACE, 2016; Belgian Science Policy, 2017; Blau, 2011; Commission of the European Communities, 2003; Dempsey, 2003; European Council/Council of the European Union, 2017; EU4Journalists, 2011; Krige and Russo, 2000; Marta, 2013; Masson-Zwaan, 2010; Robinson and Romancov, 2014; The Hague Manifesto, 2016; Vasileiou 2013a, 2013b, 2014a, 2014b, 2015, 2017a, 2017b, 2017c, 2017d and 2018).

Following a meticulous examination of more than a few EU archives, though, we eventually stress the fact that all the abovementioned particularly intricate, but concurrently extremely rousing processes are indeed well-designed and serviceable.

Special emphasis is placed on our ascertainment that a further dominant aspiration of the Council is to constructively contribute towards a decent improvement in terms of the (always critical) business environment[14] (Commission of the European Communities, 2003; European Commission, 2016; European Council/Council of the European Union, 2017; Vasileiou, 2013a and 2014a).

In order to satisfactorily accomplish such a notable achievement, the Council practically co-legislates on detailed measures for the aforementioned enterprises in terms of a) the systematic promotion of innovation, b) a sufficient improvement regarding access to funding and c) red tape cutting. These

---

14. Especially with regard to SMEs.

tribulations keep torturing many people and urgently call for a rapid response (Commission of the European Communities, 2003; European Commission, 2016; European Council/Council of the European Union, 2017; Vasileiou, 2013a and 2014a).

Our analysis carries on with regard to the multifunctional role of the European Commission. In general, space industry indeed contributes to plentiful Europe 2020 Strategy objectives for smart, sustainable and inclusive growth (Commission of the European Communities, 2003; Copernicus, 2017; EGNOS, 2017; Europa/Space, 2017; European Commission, 2016; European Commission, 2017; Galileo, 2017; GNSS, 2016· GSA/About GSA, 2015; GSA/What we do, 2016; Marta, 2013; Procaccia and Sido, 2012).

By efficaciously supporting and methodically promoting scientific progress, space industry provides the Union with abundant remarkable benefits, while simultaneously endeavouring to an incessant scuffle against unemployment.

The sectors most affected are navigation, telecommunications and Earth observation. Progress in these areas results not only in a higher degree of security and independence for the Union in terms of space policy, but also in adequately addressing numerous burning issues[15]. Over and above that, space industry, through its multifaceted systems and services, offers the EU strategically important knowledge with the

---

15. *Such as health, climate change, ageing population and scarce resources.*

aim of sufficiently supporting its external relations[16] (ASD-EUROSPACE, 2016; Commission of the European Communities, 2003; Copernicus, 2017; EGNOS, 2017; Europa/Space, 2017; European Commission, 2016; European Commission, 2017; EU4Journalists, 2011; Galileo, 2017; GNSS, 2016; GSA/About GSA, 2015; GSA/What we do, 2016; Marta, 2013; Procaccia and Sido, 2012; Vasileiou 2013a, 2013b, 2014a, 2014b, 2015, 2017a, 2017b, 2017c, 2017d and 2018).

It is obvious that the foremost target of space industry is to substantially strengthen the EU and therefore it must be regarded as a key driving force towards growth, increased competitiveness and a significant improvement in terms of quality of life all over the Union.

Consequently, the overarching priorities of a rational policy in the field of the Union's space industry, according to a Commission Communication of 2013, are the following five: a) the fastest possible development of markets for space applications and services, b) the successful securing regarding the always desirable European independent access to space[17], c) a further development of a highly competitive industrial base in Europe[18], d) the rapid establishment of a sufficient regulatory framework in order for space-related activities to be satisfactorily supported and e) an adequate assistance

---

16. Particularly within the context of development assistance and humanitarian aid.
17. As well as a gradual limitation or, if possible, a total elimination of the Union's technological dependence.
18. Accompanied by a considerable degree of encouragement in terms of the SMEs participation within the sector.

in order for the global competitiveness of the Union's space industry to be efficaciously maintained[19] (Commission of the European Communities, 2003; Copernicus, 2017; EGNOS, 2017; Europa/Space, 2017; European Commission, 2016; European Commission, 2017; Galileo, 2017; GNSS, 2016; GSA/About GSA, 2015; GSA/What we do, 2016; Marta, 2013; Procaccia and Sido, 2012).

Beyond doubt, the commitment to the creation of an appropriate regulatory framework for the Internal Market in space products, applications and services is one of the most decisive EU policy pillars within the space industry context (Commission of the European Communities, 2003; Copernicus, 2017; EGNOS, 2017; Europa/Space, 2017; European Commission, 2016; European Commission, 2017; Galileo, 2017; GNSS, 2016; GSA/About GSA, 2015; GSA/What we do, 2016; Marta, 2013; Procaccia and Sido, 2012; Vasileiou 2013a, 2013b, 2014a, 2014b, 2015, 2017a, 2017b, 2017c, 2017d and 2018). It is essential for The Proposal for a Directive on the dissemination of Earth observation satellite date for commercial purposes [COM (2014) 344] to be highlighted, due to the fact that it is an unparalleled example of a relatively recent, as well as highly significant legislative initiative. Its dominant aim is to ensure a far more effective access to high resolution earth observation satellite data (HRSD) (European Commission, 2016; European

---

19. *By properly aiding the sector so as to make it far more cost-efficient.*

Commission, 2017; Vasileiou 2013a, 2013b, 2014a, 2014b, 2015, 2017a, 2017b, 2017c, 2017d and 2018).

In tandem with HRSD-based applications, the Proposal is a first-rate tool serving a variety of key issues, such as a) the rational management of natural resources, b) defense and security, c) urban planning, d) agriculture, e) environment monitoring and f) disaster and emergency management (European Commission, 2016; European Commission, 2017; Vasileiou 2013a, 2013b, 2014a, 2014b, 2015, 2017a, 2017b, 2017c, 2017d and 2018).

According to our research, for the time being, regulations governing commercial activities using HRSD differ between EU member states. Alas, this fact results in numerous (and occasionally insurmountable) barriers to further market development, since it often hinders access to specific business-critical data (Copernicus, 2017; EGNOS, 2017; Europa/Space, 2017; European Commission, 2016; European Commission, 2017; Galileo, 2017; GNSS, 2016; GSA/About GSA, 2015; GSA/What we do, 2016; Marta, 2013).

Such a frustrating phenomenon calls for particular attention and immediate confrontation, due to the fact that its consequences are extremely complicated.

Hence, the aforesaid proposed Directive will seek to provide a functional facility for our planet's commercial observation and an effective satellite data access within the Union. A foremost pursuit is the sector's sufficient development, coupled with the successful creation of new products and services (Copernicus, 2017; EGNOS, 2017; Europa/Space, 2017; European Commission, 2017; Galileo, 2017; GNSS, 2016; GSA/About GSA, 2015; GSA/What we do, 2016; Marta, 2013; Procaccia and Sido, 2012).

It is highly probable that the entire spectrum of the aforesaid objectives will eventually become a reality by introducing a) a common definition of the HRSD concept[20], b) common standards for transparency, evenhandedness, legal certainty and predictability and c) common standards in terms of a satisfactory degree of efficaciousness combined with the tremendously important business-friendly implementation[21] (European Commission, 2016; European Commission, 2017).

A further fundamental concept is that of "Standardization", which is of skyscraping noteworthiness in terms of ensuring the most productive use of space-based systems and the opening-up of new markets for space-based services. Clear standards incontrovertibly contribute a great deal towards the definition of type and shape of future markets. Additionally, they create a stable basis for investment decisions[22]. The latter is of utmost noteworthiness, especially while experiencing economic and financial crises (European Commission, 2016; European

---

20. *Clarifying in a methodical way which satellite data can be regarded as "high resolution" ones and actually need regulation and which data are already ready for business.*
21. *With special emphasis on member states precise procedures in order to regulate the enormously notable HRSD dissemination.*
22. *We have to add that in several cases, public bodies make frequent use of space systems and therefore have the opportunity to further accelerate the functional development of standards, which is a tremendously important factor.*

Commission, 2017).

Investment uncertainty can be characterized as a wild thorn that makes every effort to economic progress unimaginable both for the present and the future. Therefore, it is obligatory to combat it at any cost.

Moreover, European standardization organizations, namely the European Telecommunications Standards Institute, the "European Committee for Standardisation" and the "European Committee for Electrotechnical Standardisation" have been mandated by the Commission to develop standards for the space manufacturing and service industries based on the work already carried out by the "European Cooperation for Space Standards Organisation" (European Commission, 2016; European Commission, 2017; EU4Journalists, 2011; Procaccia and Sido, 2012; Vasileiou, 2015, 2017a and 2017b).

For the moment, there is an immediate necessity for full interoperability between national and European space and ground-based systems. Such an aspect must be meticulously taken into consideration in order for possible future malfunctions to be avoided (European Commission, 2016; European Commission, 2017; EU4Journalists, 2011; Procaccia and Sido, 2012).

Finally, the role of SMEs in strengthening the competitiveness of the European space manufacturing industry is immensely crucial, due to the fact that they are indubitably vital for the rapid development of numerous downstream services and applications. Nowadays, the European satellite navigation and Earth observation service industry are principally made up of SMEs and start-ups (ASD-EUROSPACE, 2016; Commission of the European Communities, 2003; Copernicus, 2017; EGNOS, 2017; Europa/Space, 2017; European Commission, 2016; European Commission, 2017; Galileo, 2017; GNSS, 2016; GSA/About GSA, 2015; GSA/What we do, 2016; Marta, 2013; Procaccia and Sido, 2012).

The Section for the Single Market, Production and Consumption (INT) of the European Economic and Social Committee plays its own notable role in rationalizing the preparation of opinions on the European Economic and Social Committee a) on its own initiative, b) at the request from the European Commission, the European Parliament and the Council of Ministers and c) at the request of EU Presidencies in office (European Commission, 2016; European Economic and Social Committee, 2017).

From its name alone, it becomes apparent that the Section's dominant concern is the fertile integration of the Single Market, accompanied by the latter's smooth functioning. Such a task includes market policies, industrial policy[23], intellectual property, consumer protection, customs union, competition policy, research, company law, professions, SMEs, social economy[24] and services[25] (European Commission, 2016;

---

23. *Both general and sectoral.*

European Economic and Social Committee, 2017; Procaccia and Sido, 2012; Vasileiou, 2013a, 2013b, 2014a and 2017b).

As soon as burning issues arise, the Section arranges public hearings, in order to consult civil society organizations and it is responsible for the particularly important "European Consumer Day". Furthermore, it possesses a noteworthy Single Market Observatory (SMO), in charge of monitoring the Single Market's functions in general and proposing constructive solutions to problems that may arise. The Observatory was set up in 1994, enjoying full support of EU institutions (European Commission, 2016; European Economic and Social Committee, 2017).

The Commission for the Environment, Climate Change and Energy (ENVE) performs more than a few productive activities by coordinating in a well-functioning manner the Committee of the Regions' work on a) space policy for territorial development, b) climate change[26], c) environmental policy, d) renewable energy, e) new energy policies and f) trans-European networks[27] (CoR Commissions, 2017; European Commission, 2016; Procaccia and Sido, 2012; Vasileiou, 2014b and 2017a).

Apart from the abovementioned institutions, an assortment of key agencies, such as the ESA, the GSA, the EEA and the "EU Satellite Centre", are actively involved in the broader context of the EU space policy. Needless to say, their significance is no

---

24. Comprising mutual societies, associations, foundations and cooperatives.
25. Including trade, tourism, insurance and the banking sector. We have to mention, though, that utilities are not included.
26. Adaptation and mitigation.
27. In the context of the energy sector.

less than paramount.

The majority of the ESA's principal actions and competencies have already been addressed, but a number of additional elements for further knowledge are indeed worth mentioning.

The ESA is an intergovernmental agency run by 22 countries (Austria, Belgium, Czech Republic, Denmark, Estonia, Finland, France, Germany, Greece, Hungary, Ireland, Italy, Luxembourg, the Netherlands, Norway, Poland, Portugal, Romania, Spain, Sweden, Switzerland and the United Kingdom). Slovenia is an Associate Member, while attention must be paid to the fact that Canada is also involved in certain projects under a cooperation agreement (Belgian Science Policy, 2017; Blau, 2011; Dempsey, 2003; Europa/Space, 2017; Krige and Russo, 2000; Marta, 2013; Masson-Zwaan, 2010; The Hague Manifesto, 2016).

We strongly emphasize the fact that Cyprus, Malta, Bulgaria, Slovakia, Latvia and Lithuania also have cooperation agreements with the ESA. The latter's headquarters are located in Paris and we add that the ESA has been established following the decision to merge the two pre-existing space policies' European organizations, namely the "European Launcher Development Organisation" (ELDO) and the "European Space Research Organisation" (ESRO) (Belgian Science Policy, 2017; Blau, 2011; Dempsey, 2003; Europa/Space, 2017; Krige and Russo, 2000; Marta, 2013; Masson-Zwaan, 2010; The Hague Manifesto, 2016).

In order to proceed to a more scrupulous review of the GSA, it is essential to clarify the precise concept of GNSS, alongside with its specific characteristics and peculiarities.

GNSS is a satellite constellation providing space signals which transmit positioning and timing data to GNSS receivers. The latter then use these data in order to successfully determine location[28] (GNSS, 2016; GSA/About GSA, 2015; GSA/What we do, 2016).

GNSS performance is an utterly cardinal parameter for dozens of important functions and is systematically evaluated through four pivotal criteria. The first is continuity, which is the capacity of the system to uninterruptedly operate. We highlight the fact that in most cases interruptions prove to be absolutely detrimental (GNSS, 2016; GSA/About GSA, 2015; GSA/What we do, 2016).

The second is accuracy, interpreted as the precise difference between a receiver's measured and real position, speed or time. The third is integrity, which is the ability of the system to provide a threshold of confidence and to trigger an alarm in case of problems regarding positioning data, while the fourth is availability, explained as the percentage of time a signal fulfils the three aforesaid criteria[29] (GNSS, 2016; GSA/About GSA, 2015; GSA/What we do, 2016; Procaccia and Sido, 2012).

Hence, we state that the GSA's dominant target is the prolific support of the Union's objectives, accompanied

---

28. As it has already been noted, examples of GNSS include Galileo, GPS, GLONASS and BeiDou.
29. It must also be mentioned that the overall performance of GNSS can be considerably improved by SBAS such as EGNOS, which contributes in a decisive manner to the palpable advancement of GPS information.

by the achievement of the highest possible return on European GNSS investment, with regard to benefits for users, economic growth and competitiveness by a) guaranteeing that European GNSS services, as well as operations are indeed safe and accessible, b) adequately managing to ceaselessly provide users with the best possible services[30], c) designing and enabling services that successfully meet user needs[31] and d) sufficiently engaging market stakeholders to create innovative applications, value-added services and user technology that unremittingly promote the coveted achievement of the full European GNSS adoption in a methodical way (GNSS, 2016; GSA/About GSA, 2015; GSA/What we do, 2016; Procaccia and Sido, 2012).

In other words, the GSA is constantly attempting to establish a functional connection between space technology and its users. We obstinately argue that its foremost target is the direct "translation" of Galileo and EGNOS signals into credible services for EU citizens. Emphasis must be placed on the fact that the Commission has entrusted the GSA with the responsibility regarding not only the Galileo service operations, but also the latter's initial services (ASD-EUROSPACE, 2016; EU4Journalists, 2011; GNSS, 2016; GSA/About GSA, 2015; GSA/What we do, 2016; Procaccia and Sido, 2012; Robinson and Romancov, 2014).

The GSA is literally "obliged" to remain in incessant contact with industry, stakeholders and user communities and conduct

---

30. *Always in a cost-efficient manner.*
31. *Combined with the uninterrupted improvement of European GNSS services and infrastructure.*

a constructive dialogue on a vast array of neuralgic issues. More specifically, it has to work closely with chipset and receiver manufacturers to ensure that all products are compatible with Galileo and can actively contribute towards its further rational development and exploitation (GNSS, 2016; GSA/About GSA, 2015; GSA/What we do, 2016; Procaccia and Sido, 2012; Robinson and Romancov, 2014; Vasileiou, 2013a, 2014a, 2017a and 2017b).

Besides, it unremittingly cooperates with rail and maritime stakeholders, so that they can properly upgrade their systems to be ready to sufficiently use Galileo. Additionally, it is worth highlighting that programs such as Horizon 2020, offer notable assistance in order for the required readiness level to be precisely reached (GNSS, 2016; GSA/About GSA, 2015; GSA/What we do, 2016; Procaccia and Sido, 2012; Robinson and Romancov, 2014; Vasileiou, 2013a, 2014a, 2017a and 2017b).

We have thus far drawn the conclusion that the GSA has indeed managed to provide significant support in terms of the successful promotion regarding the use of EGNOS, always for the general interest. In particular, at present, more than 200 airports operate with EGNOS-based approaches and over two-thirds of European tractors actually benefit from "precision farming", which relies on EGNOS. Apart from that, EGNOS is the standard for mapping, as well as surveying in Europe (GNSS, 2016; GSA/About GSA, 2015; GSA/What we do, 2016; Robinson and Romancov, 2014).

The EEA's primary objective, as its name implies, is the provision of direct, valid and independent information concerning the environment. We remain certain that the EEA is one of the the most significant sources of information, not only for those actively involved in the development, adoption, implementation and evaluation of environmental policy, but also for the general public (European Environment Agency, 2016; Vasileiou, 2014b).

In 1990, the regulation for the establishment of the EEA was adopted by the Union, and three years later[32] it came into force, following the decision for the EEA to be located in Copenhagen. More intense work began in 1994 and we add that the regulation actually led to the creation of the European environment information and observation network (Eionet) (European Environment Agency, 2016; Vasileiou, 2014b).

In short, the EEA's mandate is: a) to advantageously assist the Community, as well as member and cooperating countries in the decision-making process[33] and b) the sufficient coordination of Eionet, which is a rather complicated and time-consuming task. Currently, the EEA comprises 33 member countries (EU member states, Switzerland, Turkey, Norway, Iceland and Liechtenstein) and six cooperating ones (Albania, Bosnia and Herzegovina, FYROM, Serbia, Montenegro and Kosovo on the basis of United Nations Security Council Resolution 1244/99)

---
32. At the end of 1993.
33. In terms of environmental improvement progress towards sustainability and the productive integration of the "environmental considerations" into the context of economic policies.

(European Environment Agency, 2016; Vasileiou, 2014b).

Eionet is a partnership network of the countries and the EEA. The latter is in charge of the rational development of the network, as well as the sufficient coordination of its activities. That is why it works closely with "national focal points"[34] responsible for the constructive coordination of national networks, where approximately 350 institutions are actually involved (European Environment Agency, 2016; Vasileiou, 2014b).

According to the EU, the foremost clients are the Commission, the Council, the Parliament, as well as member and cooperating countries, while the Committee of the Regions and the Economic and Social Committee are also being served. Furthermore, it must be noted that users of the EEA information are academics, businessmen, non-governmental organizations and miscellaneous civil society parts and members (European Environment Agency, 2016; Vasileiou, 2014b).

The "EU Satellite Centre" is located in Torrejón de Ardoz, in the vicinity of Madrid. It was established in 1992, and a decade later, namely on January 1, 2002, it was successfully incorporated as an agency into the Union (Robinson and Romancov, 2014; SatCen, 2017; Vasileiou, 2015 and 2017a).

Its role is undeniably crucial since it actively supports the Union's decision-making process in the complex (and often risky) field of the Common Foreign and Security Policy (CFSP) and in

---

34. *We mainly mean environment ministries or national environment agencies.*

particular as regards the Common Security and Defense Policy (CSDP), including EU crisis management missions (Robinson and Romancov, 2014; SatCen, 2017; Vasileiou, 2015 and 2017a).

Such a priceless backing becomes a reality through the satisfactory provision of products and services resulting from the rational exploitation of both relevant space assets and collateral data including satellite/aerial imagery, as well as other equally noteworthy related services (Robinson and Romancov, 2014; SatCen, 2017; Vasileiou, 2015 and 2017a).

We irrefutably claim that the EU has thus far been successful in the context of its space policy, mainly due to the fact that it manages to achieve a satisfactory independent access to space at a reasonable cost. Such an accomplishment, in the majority of cases, can effectively be regarded as the recipe for success.

A truly pivotal factor, in order for the Union to remain at the forefront of the space sector is the incessant improvement of its industrial competitiveness. Nobody can deny that the EU industry has already been competitive enough and massive unremitting efforts indeed take place towards further progress and development.

What we attempt to clarify, though, is that due to the relentless competition (which is highly probable to intensify in the future), it is essential for the Union to steadily continue its attempts for further improvement and to be able to anticipate as accurately as possible future needs.

An additional critical issue that arises and in quite a lot of cases monopolizes relevant discussions is whether the Union can actually increase spending on space policy.

It goes without saying that it is utterly impossible for such a question to be answered in a few lines. Be that as it may, we strongly believe that the economic crisis, accompanied by all inevitable budget constraints must be seriously taken into account in order for an array of fruitful conclusions to be effectively drawn.

We certainly welcome the view that, precisely due to the political and economic instability and the immense cost of space programs and strategies, the Union must continue operating by taking careful steps and by critically observing developments that in the majority of cases change rapidly.

Indubitably, a functional space policy provides enormous benefits to the community, but both the design and practice are always dictated by the existing funds. Therefore, the admittedly tricky challenge that the Union is called upon to face is the achievement of a rationalized space policy, amidst the current economic obscurity.

In the following (and final) chapter, a systematic outline of the foremost conclusions regarding the efficiency of the Union's space policy takes place. We attempt to highlight all focal points, evaluate the current degree of success, identify the major problems and point out the most important future prospects.

CHAPTER 3

CONCLUDING REMARKS

A critical analysis of the Union's space policy and the way it affects our everyday life has been our principal concern while authoring this book. In the previous chapters, numerous objectives, difficulties and prospects related to space policy have been methodically examined, and, concurrently, several key sectors, as well as pivotal institutions and agencies have been painstakingly analyzed.

As it has already been noted, the majority of services on our planet, such as television, telecommunications, financial systems and meteorological forecasts, are directly dependent on the space sector and, by extension, on the level of technologies, discoveries and inventions associated with it.

Furthermore, it has been underlined that the space sector irrefutably possesses the ability to actively contribute to numerous efforts towards the desirable resolution of abundant thorny global problems, the treatment of which is considered to be more than urgent. Subsequently, in the context of our attempt to constructively summarize the foremost points examined within the book's main body, an abundance of fruitful conclusions can be effectively drawn. The latter may serve as a functional guide to future developments and relevant scientific quests that might pave the way for an even higher degree of growth and progress. Data arising from the rational exploitation of space and the flourishing pursuit of space policy can be successfully implemented in the context of climate change, health, unemployment, defense, security, energy management, meteorology, monitoring environment and the marine environment, environmental protection, natural resource management, industrial competitiveness, aviation, maritime navigation, railways, agriculture, biotechnology recycling, forestry, disaster response, risk signs, rescue operations, civil protection, crises or emergencies, traffic, transport, urban planning and tourism.

The principal areas of the Union's space policy are Copernicus system, Galileo and EGNOS satellite programs, space exploration, and space research. Following an exhaustive scrutiny of numerous EU archives, we have indeed acknowledged that all four have already offered a great deal to the benefit of the entire globe.

Moreover, in order for a smooth and prolific space policy practice to take place, a top-class cooperation combined with a first-rate coordination between the European Parliament, the Council of the EU, the European Commission, the European Economic and Social Committee, the Committee of the Regions, the ESA, the GSA, the EEA and the EU Satellite Center, is literally indispensable.

We have repeatedly stressed the fact that the above spectrum is enormously broad and outstandingly complex. Nonetheless, we imagine that through our analysis, the reader has already increased his level of knowledge in terms of the structure, the functions and the Union's general approach regarding space and its peculiarities.

We deliberately refrained from using strict scientific terminology and avoided entering into technical details, which could well be cited, especially in the description of Copernicus, Galileo and of EGNOS. Such a decision stems from the fact that the current book is not intended to substitute for space research or satellite technology textbooks.

On the contrary, its purpose is to critically examine the majority of issues that connect the Union with space in a comprehensible way for all readers, regardless of scientific training. We remain optimistic that this book may well serve as a compass for further research, using simple language, but without omitting pivotal details and parameters.

We believe that the bibliography provided is sufficient enough for the reader to a) further carry on with his personal analysis, b) critically delve into his preferred issues, c) point out any possible misinterpretations and d) predict (if possible) any future challenges and strategies.

Each book is considered to be just an intermediate point in the ongoing flow of science and it is beyond doubt that careful recording, accompanied by punctilious analysis regarding future developments will serve as the roadmap for greater advancement and more innovative technological findings.

The issues of both space policy and space research in general can actually be characterized as inexhaustible. The brilliant unremitting scientific progress, coupled with new discoveries that constantly come to light, clearly indicate the wealth of knowledge humanity can efficaciously acquire via systematic and ceaseless research.

We have hitherto acknowledged that the EU space policy does not affect only the Union and its citizens. In contrast, it is of utmost noteworthiness for the entire globe and taking this parameter meticulously into consideration, we are indeed confident that strengthening the competitiveness of EU space industry can eventually result in a more positive universal impact both for the present and the future.

The first timid, but at the same time tremendously ambitious steps for space conquest, which took place in the middle of the last century, have been substituted by present modern means. Space science has progressed to such an extent that we assert without exaggeration that almost every day something new comes out. Therefore, there is no question as to whether new discoveries come to light, but if the latter are actually subjected to rational operation and well-regulated management for the general interest as such.

We unquestionably support the fact (and we imagine that the reader will agree) that up to now the Union's space policy has been satisfactorily conducted and has sufficiently contributed towards serving the general interest in the best possible manner.

Alas, the dizzying speed of economic, social and political conditions makes any accurate future forecasts tremendously problematic, if not utterly impossible. Based on background data and general guidelines for future policy designing, it is undeniably possible to somehow anticipate a number of developments, but absolute accuracy is totally unfeasible.

Naturally, we do not intend to express fears about the future in any way, and this must be made clear. Time, as always, is the only (relentless) judge. What we would like to elucidate, though, is the fact that, following loads of hard struggles, the Union has finally succeeded in adequately putting space into everyday life's service.

2020 is indeed too close, so then we will be in position to proceed to a more scrupulous assessment of a) the already achieved objectives, b) existing problems, c) the degree of success regarding strategies and policies, d) potential dysfunctions and e) future challenges and prospects.

We remain positive that data that will emerge in 2020 will be outstandingly crucial. Taking the opinion of numerous EU experts meticulously into account, the entire range of such data will deeply influence the space policy design for the coming decades.

According to the general philosophy of the Union, only the best is good enough for a brighter future. Consequently, a vast array of systematic attempts to discover new paths of growth and development will be (as usual) the most imperative EU pursuit for a more constructive present and a more optimistic tomorrow.

BIBLIOGRAPHY

ASD-EUROSPACE (2016), "*A Space Strategy for Europe-Contribution of the European space industry*", Position Paper, available at http://www.eurospace.org/Data/Sites/1/eurospacepositionpaper_spacestrategy.pdf (accessed on 26/3/17).

Belgian Science Policy (2017), "*Europe: ESA-membership*", available at https://www.belspo.be/belspo/space/euPolicy_esa_en.stm (accessed on 15/3/17).

Blau, John (2011), "*The European Union Shoots for the Stars*", Research Technology Management, 54 (4), pp. 3-5.

Commission of the European Communities (2003), "*Green Paper/European Space Policy*", Brussels, 21.1.2003 COM(2003) 17 final, available at http://www.globalsecurity.org/space/library/policy/int/eu_greenpaper_01jan2003.pdf (accessed on 26/3/17).

Copernicus (2017), "*Copernicus-The European Earth Observation Programme*", available at http://ec.europa.eu/growth/sectors/space/copernicus/ (accessed on 12/3/17).

CoR Commissions (2017), "*CoR Commissions-Commission for the Environment, Climate Change and Energy (ENVE)*" (in Greek), available at http://cor.europa.eu/el/activities/commissions/Pages/cor-commissions.aspx?comm=ENVE (accessed on 28/2/17).

Dempsey, Judy (2003), "*EU 'must double' space spending*", Financial Times, (11/11/2003), London, UK.

EGNOS (2017), "*EGNOS*", available at http://ec.europa.eu/growth/sectors/space/egnos/ (accessed on 22/2/17).

Europa/Space (2017), "*Space*" (in Greek), available at https://europa.eu/european-union/topics/space_el (accessed on 22/2/17).

BIBLIOGRAPHY

European Commission (2014), *"The European Union Explained: Enterprise" (in Greek), available at https://europa.eu/european-union/topics/ space_el (the document is included in this webpage) (accessed on 15/3/17).*

European Commission (2016), *"Communication from the Commission to the European Parliament, the Council, the European Economic and Social Committee and the Committee of the Regions-Space Strategy for Europe", Brussels 26.10.2016 COM(2016) 705 final, available at https://ec.europa.eu/transparency/regdoc/rep/1/2016/ EN/COM-2016-705-F1-EN-MAIN.PDF (accessed on 31/3/17).*

European Commission (2017), *"The space industry", available at http:// ec.europa.eu/growth/sectors/space/industry_el (accessed on 28/2/17).*

European Council/Council of the European Union (2017), *"Competitiveness Council (COMPET)" (in Greek), available at http://www.consilium.europa.eu/el/council-eu/ configurations/compet/ (accessed on 28/2/17).*

European Economic and Social Committee (2017), *"Single Market, Production and Consumption (INT)", available at http:// www.eesc.europa.eu/?i=portal.en.int-section (accessed on 28/2/17).*

European Environment Agency (2016), *"Who we are" (in Greek), available at http://www.eea.europa.eu/el/about-us/who (accessed on 6/3/17).*

European Parliament (2017), *"European Parliament-Committees-Industry, Research and Energy (ITRE)" (in Greek), available at http://www. europarl.europa.eu/committees/el/itre/home.html (accessed on 2/3/17).*

European Parliament/Annex V: Powers and Responsibilities of Standing Committees (2017), *"IX. Committee on Industry, Research and Energy" (in Greek), available at http://www.europarl.europa.eu/sides/ getLastRules.do?language=el&reference=RESP-ITRE (accessed on 2/3/17).*

BIBLIOGRAPHY

EU4Journalists (2011), "Space Policy/1. New on the EU agenda: European space policy", eu4journalists Dossiers, available at http://www.eu4journalists.com/index.php/dossiers/english/C79/index.html# (accessed on 23/3/17).

Galileo (2017), "Galileo", available at http://ec.europa.eu/growth/sectors/space/galileo/ (accessed on 22/2/17).

GNSS (2016), "What is GNSS?", available at https://www.gsa.europa.eu/european-gnss/what-gnss (accessed on 6/3/17).

GSA/About GSA (2015), "About GSA", available at https://www.gsa.europa.eu/gsa/about-gsa (accessed on 6/3/17).

GSA/What we do (2016), "What we do", available at https://www.gsa.europa.eu/about/what-we-do (accessed on 6/3/17).

HORIZON 2020/Leadership in Enabling and Industrial Technologies (2017), "HORIZON 2020-The EU Framework Programme for Research and Innovation-Leadership in Enabling and Industrial Technologies", available at http://ec.europa.eu/programmes/horizon2020/en/h2020-section/leadership-enabling-and-industrial-technologies (accessed on 22/2/17).

Krige, J. and Russo, A. (2000), "A History of the European Space Agency 1958-1987, Volume 1: The story of ESRO and ELDO, 1958-1973" (with contributions by M. De Maria and L. Sebesta), ESA SP-1235, Απρίλιος 2000, available at http://www.esa.int/esapub/sp/sp1235/sp1235v1web.pdf (accessed on 12/3/17).

Marta, Lucia (2013), "Consolidating the European Space Policy requires an evolution of its governance, which is currently structured around three main types of actor", EUROPP (European Politics and Policy), The London School of Economics and Political Science, available at http://blogs.lse.ac.uk/europpblog/2013/06/21/european-

BIBLIOGRAPHY

space-policy-is-governed-by-a-triangle-of-the-eu-the-european-space-agency-and-national-space-agencies/ (accessed on 12/3/17).

Masson-Zwaan, Tanja (2010), "Recent Developments in EU Space Policy and Law", Leiden University, available at https://openaccess.leidenuniv.nl/bitstream/handle/1887/17593/Tanja_Masson-Zwaan_paper.pdf?sequence=1 (accessed on 11/3/17).

Procaccia, Catherine and Sido, Bruno (2012), "European Space Policy: A time for decisions", Report on the Challenges and Prospects for European Space Policy, The French Republic, Parliamentary Office for the Evaluation of Scientific and Technical Choices, available at https://www.senat.fr/fileadmin/Fichiers/Images/opecst/quatre_pages_anglais/4p_recomman_polit_spatiale_2012_2.pdf (accessed on 23/3/17).

Robinson, Jana and Romancov, Michael (2014), "The European Union and Space: Opportunities and Risks", Non-Proliferation Papers 37, EU Non-Proliferation Consortium, available at http://www.nonproliferation.eu/web/documents/nonproliferationpapers/janarobinsonmichaelromancov52e8eeb2de1b1.pdf (accessed on 26/3/17).

SatCen (2017), "The Centre", available at https://www.eusc.europa.eu/about_the_eu_satcen/the_centre (accessed on 6/3/17).

The Hague Manifesto (2016), "The Hague Manifesto on Space Policy", The Hague, June 2016, The Netherlands Presidency of the Council of the European Union, available at http://www.gsrt.gr/News/Files/New11266/The%20Hague%20Manifesto%20on%20Space%20Policy.pdf (accessed on 15/3/17).

Vasileiou, Ioannis (2013a), European Unification-A Process of Convergence, or Divergence? (in Greek) (Athens: Historical Quest).

Vasileiou, Ioannis (2013b), "1980-1999, European Union: The Years of Expansion and Enlargement", From Hitler's New Europe to Merkel's

BIBLIOGRAPHY

Eurozone (in Greek), Vol. 1, Historical Archive of Ependytis, pp. 76-95.

Vasileiou, Ioannis (2014a), *European Unification-A Process of Convergence, or Divergence? (2nd Edition-Special Edition for Universities) (in Greek) (Athens: Historical Quest).*

Vasileiou, Ioannis (2014b), *The Present and Future of the Agricultural Policy of the European Union (in Greek) (Athens: Historical Quest).*

Vasileiou, Ioannis (2015), *The Foreign and Security Policy of the European Union-A Critical Approach (in Greek) (Athens: Historical Quest).*

Vasileiou, Ioannis (2017a), *European Union and Energy-The Route Towards 2050-Thoughts, Ideas and Conclusions (in Greek) (Athens: Historical Quest).*

Vasileiou, Ioannis (2017b), *EU Budget-Issues about the Allocation and Redistribution of Resources in the EU (in Greek) (Athens: Historical Quest).*

Vasileiou, Ioannis (2017c), *Climate Change: Manageable Problem, or Slow Death of the Planet? Role and Actions of the EU Until 2050-The Consequences on Greece (in Greek) (Athens: Historical Quest).*

Vasileiou, Ioannis (2017d), *Economic Crisis, Employment and Social Affairs in the European Union-Proposals and Actions to Combat Unemployment (in Greek) (Athens: Historical Quest).*

Vasileiou, Ioannis (2018), *Enterprises in the EU: Monopolies-Cartels-State Aid-Competition Rules (in Greek) (Athens: Historical Quest).*

Vereshchetin, V.S. (2010), *"The Law of Outer Space in the General Legal Field (Commonality and Particularities)", Revista Brasileira de Direito Aeronáutico e Espacial, available at http://www.sbda.org.br/revista/1826.pdf (accessed on 12/3/17).*

IOANNIS VASILEIOU

BIOGRAPHY

Ioannis Vasileiou was born in Athens in 1978. In 2001, he was awarded his Ptychio (equivalent to Bachelor's degree) in Political Science and Public Administration from the University of Athens (Greece). In 2003, he was awarded his first Master's degree (International Political Economy) from the University of Warwick (UK). In 2005, he was awarded his second Master's degree (International Economic Management) from the University of Birmingham (UK). In 2011, he was awarded his PhD from the University of Birmingham (UK) with specialization in the economic and political aspects of the European Union's Regional Policy. Since 2011, he has been conducting academic research on issues related to the European Union and international politics and economics.

www.ingramcontent.com/pod-product-compliance
Lightning Source LLC
Chambersburg PA
CBHW071109240526
45469CB00006BD/2401